On this book you'll find a practical guide on how to use zoom on every device you own as well as the different tools the platform offers to help you connect with your loved ones or for video conferencing at work.

I hope you enjoy and learn with this material that was put together with the utmost care so that you can use zoom to the fullest and stay connected.

I0480520

DOUGLAS MORALES

INDEX

WHAT IS ZOOM?

Zoom is a video conferencing app which can be used to meet virtually with people all over the world whether it is through video or just audio while being able to chat with the members of the call. This call also allows you to record the sessions on your computer or the cloud.

If you work from home you have probably heard of this app or even used it at some point, right now it is one of the top video conferencing apps in the market.

When you talk about zoom, you generally hear key words such as Zoom Room and Zoom Meetings. A Zoom Meeting is a meeting held in the Zoom app and people can join through a pc, notebook, tablets and/or phones.

Zoom Room is a software configuration that allows you to schedule and start a Zoom Meeting from a conference room.

ZOOM CHARACTERISTICS

Zoom is an app that can be used with different characteristics to the plan you have chosen in the moment you download and install the app. The general characteristics Zoom has can be enumerated in 3:

1. The most important characteristic if you use the app with a free account is that you can make individual meetings with no time limit.
2. This second characteristic, (which is the most used one) is the possibility to create group video conferences with up to 500 participants with unlimited time if you buy the Big Meeting complement, if you have a free

account, you can organize meetings of up to 100 participants with a 40 min time limit.

3. The third general characteristic can be used with any subscription plan and any time limit, "Share Screen" allows you to show all the participants your screen.

HOW DOES ZOOM WORK?

The app allows the creation of individual chat sessions, group calls, training sessions and even web seminars for internal and external audiences and global video conferences with up to 1000 participants, showing 49 videos simultaneously per screen, which can be varied by swiping on the screen to see the rest of participants.

As we know Zoom offers subscription plans with prices that vary depending on the type of plan, country and currency. Below you will a a list of options:

1. Free Plan: Allows for individual meetings with no time limitations, and meetings with up to 100 participants with a 40 min time limit and the possibility to record this sessions.
2. Zoom Pro: This plan has a price of 14,99 american dollar per month, allows the host to create a personal meeting ID and session recording, limiting the group meetings up to 24 hours.
3. Zoom Business: This plan has a price of 19,99 american dollars per month per host, there should be a minimum of 10, it allows having a personalized URL and the brand of th company, this plan offers transcriptions of the sessions recorded on the cloud and dedicated customer service.

4. Zoom Enterprise: the most advanced plan with a price of 19,99 american dollars per month, per host, with a minimum of 100, it's destined for big companies with over 1000 employees, offers unlimited cloud storage for session recording, a customer success administrator and discounts in web seminars and Zoom Rooms.

Besides this 4 subscription plans, it also offers Zoom Rooms optionally, if you wish to set up Zoom Rooms you can sing up for a free 30 day trial, once the free trial has finished it has a price of 49 american dollars per months and room subscription while the web seminars have a price of 40 american dollars per host.

DOWNLOAD ZOOM

Even though Zoom allows the connection to video conferences or calls from web it is highly recommended to download the app to obtain a friendlier interface whether it is on your desk computer or smartphone

Zoom is compatible with all types of electronic devices and operative systems that we know.

Download Zoom for iOs: https://apps.apple.com/us/app/zoom-cloud-meetings/id546505307

Download Zoom for Android: https://play.google.com/store/apps/details?id=us.zoom.videomeetings&hl=en

Download Zoom for Windows: https://zoom.us/support/download

Download Zoom for MAC: https://zoom.us/support/download

Download Zoom for LINUX: https://zoom.us/support/download

Zoom's interface has some variations on smartphones and desk computers, this mean that some buttons are placed differently, you can see this here:

App screen for desktop computers (Windows, MAC, LINUX)

App screen for smartphones (Android, iOS)

Zooms allows you to join a meeting without signing in on all the different apps and devices, it also allows to sing in using a Zoom, Google, Facebook or SSO account.

Being signed in you can start a meeting, join a meeting adding the meeting ID, share screen, mute or activate microphone, start or stop video, invite other to join the meeting, change the name that appears on screen, chat and record the meeting and save the recording in the cloud and and if it's desktop user, it can record the session in the desktop's hard drive.

From the desktop app it is also possible to create polls and share through Facebook live and more, in other words, the desktop app has more functionalities but if you have a free account, it is very similar to the mobile app.

ZOOM ON YOUR TV

It is now possible to make zoom work on your TV, but if you want to have your video call on a big screen, we will show you how to watch a Zoom meeting on your TV.

For companies, Zoom has developed Zoom Rooms but for domestic use there are many options through which you can have Zoom on your TV.

Basically, it is possible to cast your screen in two ways, wired and wireless, the advantage of connecting through wire is the connection stability and the advantage of wireless connection is that you can keep the screen close to your face.

You can connect your TV through different devices that establish said connection with your phone and if you use the desktop app, you can connect through an HDMI cable from your PC to your TV

Here you have a list of the different types of devices that allow your to cast Zoom to your TV:

1. AirPlay with Apple TV o Tv's that are compatibles with AirPlay, this is the most practical way to connect if you have an iPhone, iPad or MAC, it allows you to cast your device's screen on your TV.

For iOs devices, make sure is up to date in your phone or in your Apple TV, all devices need to be connected to the same WiFi or else it won't work, then swipe down from the top right area of your iPhone or iPad, in other devices with Touch ID you have to swipe up from the bottom up and select the option 'Duplicate Screen'

After finishing with this steps, you must press on the name of the Apple TV that shows on the list and it's done, your screen will cast on the TV, you can then connect to Zoom and start your call.

For MAC the process is the same, if this option is available, you will see the icon in the bar located in the top are of the screen and then click on 'Share Screen' and it's done, you screen will cast on your TV, you can then connect to Zoom and start your call.

2. With other brands, you can choose to cast your screen using Chromecast, this device acts as a bridge between your smartphone or computer and your TV. The Chromecast is connected to the TV via HDMI and allows you to control what you see like Netflix or Amazon Prime, it also gives you the possibility to duplicate the screen of your smartphone, Chrome browser or Chrome OS device.

With Chromecast functioning, you will need to find the casting option, a bo with the WiFi logo located on the top right corner. You will find this on your Chrome browser, smartphone or Chrome OS device, the name varies depending on the Android manufacturer, Pixel, Sony and Xiaomi use the word "Cast", Samsung calls it "Smart View", Huawei "Wireless Projection" among other names, but it will be found in the settings menu on the top right corner of the screen.

When setting up your Chromecast, make sure that your phone is connected to the same WiFi network than the Chromecast, find the casting option on your phone and search for the device to share, select your Chromecast and your smartphone's screen will appear on your TV. Open Zoom and start your meeting.

With laptops the process is the same with the only difference that you will need to cast your desktop screen instead of just your browser tabs, this is because Zoom runs on its own app and if your cast just your browser's tab you will only be able to see that on your TV and not the Zoom app.

3. To cast on your TV using Roku, the process is very similar, you may not know, but Roku is also compatible with Android, so if you have an Android smartphone or PC and a Roku device, you are all set.

Just as with step number 2, the name varies depending on the manufacturer, Cast, Smart View, Wireless Projection, etc, but most smartphones are compatible with Roku

Set up your Roku device, make sure it's connected to the same WiFi network as your smartphone, find the mirror conversion option and search for the connected devices, select the Roku you are using, confirm that you want to allow the connection with your smartphone and you will be seeing your phone's screen on your TV, open Zoom and start a meeting.

Nowadays there are many Smart Tvs that allow casting without needing an additional device, for example,

Samsung Smart TV allow this with many devices, once again, you just need to open Zoom and start a meeting.

4. To connect through an HDMI cable, the internet connection will be more estable via ethernet (direct) connection, there's no need to worry for letting the WiFi disconnect, but it also means that your device will be connected to the TV will mobility or the possibility to move the camera according to where you would like to be in the moment you are in a meeting.

Another advantage with HDMI connection is how simple it is to connect if your computer has an HDMI port, connect the cable to the port and the other end to the HDMI port on your TV and that is enough to be able to see your desktop screen on your TV, and thanks to the connection via HDMI you will be able to use the sound of your TV as well.

if you don't have an HDMI port, but you do have an USB-C cable/port, you can also connect a dongle that is compatible with HDMI, there are adapters that do not recognize HDMI connection which is why it is recommendable to have one that is made by the same manufacturer as the device you are connecting. After that, you just need to open Zoom and start a meeting.

If you wish to improve sound quality, you can connect your smartphone or computer via Bluetooth, this means you can connect a speaker and have it near you without needing to stay close to your device in order to listen better. If the speaker also has a microphone, this would allow an even better experience.

What is the difference between Zoom paid and free accounts?

There are some key differences int he variety of accounts that Zoom offers that are worth mentioning.

TIPS ON HOW TO USE ZOOM

You will learn some tips about the Zoom app to improve the video conferencing experience

1. Zoom allows you to schedule meetings, you can set up the recurrent meeting on the day and time you wish to have it and use the same URL link each time.

Desktop app screen

Smartphone app screen.

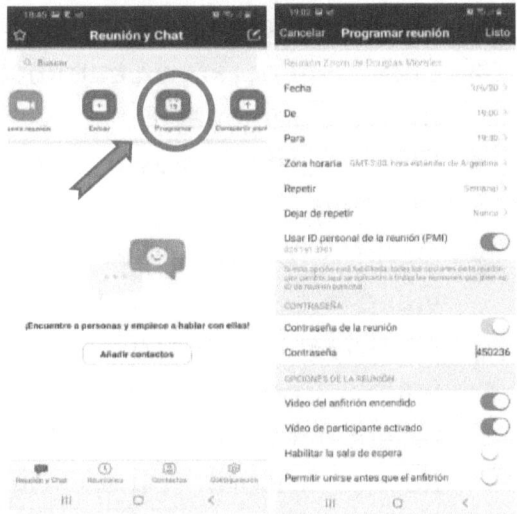

After signing in and scheduling the meeting, you can click on "repeat Schedule" and set up the parameters for your meeting

- Name of the meeting.
- Date, time and time zone.
- Don't repeat, repeat every day, weekly, every 2 weeks, monthly or annually.

- Stop repeating if there is an specific date in which you wish to stop the meetings.
- Click on generate an automatic ID or use a personal ID.
- Choose the password of the meeting.
- Options of the meeting: Host video preferences and participants (ON/OFF), allow waiting room, allow participants to enter the meeting before the host, record the session automatically.
- Add meeting to Outlook or Google Calendar or any other calendar.

2. To record the meeting, the host will need to abilitate session recording in the account configuration, after starting the meeting, click on account/meeting configuration the click on the recording tab and allow recordings. This

configuration is only for the
desktop app.

Desktop app screen.

To allow participants to record the
meeting while being on the call / video
conference, the host should click on
"Manage participants" and on the menu
click on the button next to the name of

the participant you want to allow to record and click on "allow recording"

The host can allow all participants to record meetings and choose where the session would be saved, in the cloud or hard drive. It is noteworthy to mention that recorded meetings that are saved in the cloud are only for paid accounts.

Follow these steps if you want to record from the mobile app:

- Start the meeting from the app.
- Click on the three dots that appear on right bottom corner of your screen.
- Finally, click on "record" and an recording icon will appear and you will be able to also pause said recording.
- The recording will be saved also in the section "My recordings" on Zoom.

Zoom saves the recordings locally on the Zoom folder on your PC, usually he rout to that folder is:

C:\ User \ Name of user \ Documents \ Zoom

On the other hand, you can also access easily the recordings from the app by clicking on "recordings" that appears on each meeting located on the meeting history record in the app.

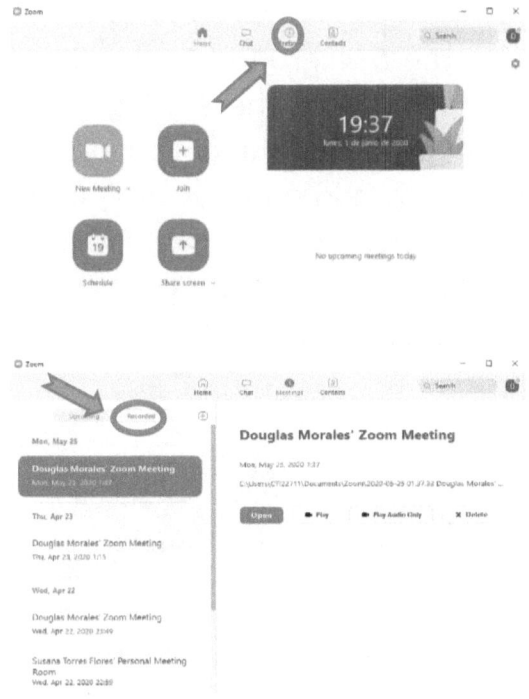

Desktop app screen.

3. If you would like to make your meetings a bit more cheerful or if you simply want to avoid people from looking at your background, you can add virtual backgrounds, Zoom offers several types of urban backgrounds, space or ocean views, but you can also upload the image you prefer.

To add the background in the desktop app, simply go to the setting tab and choose the background you prefer, if you want to add another one, click on the + simbol on the top left and choose the image you want from your pc

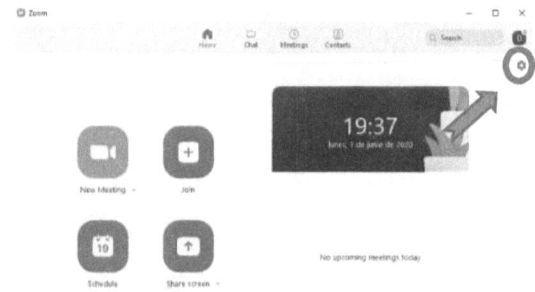

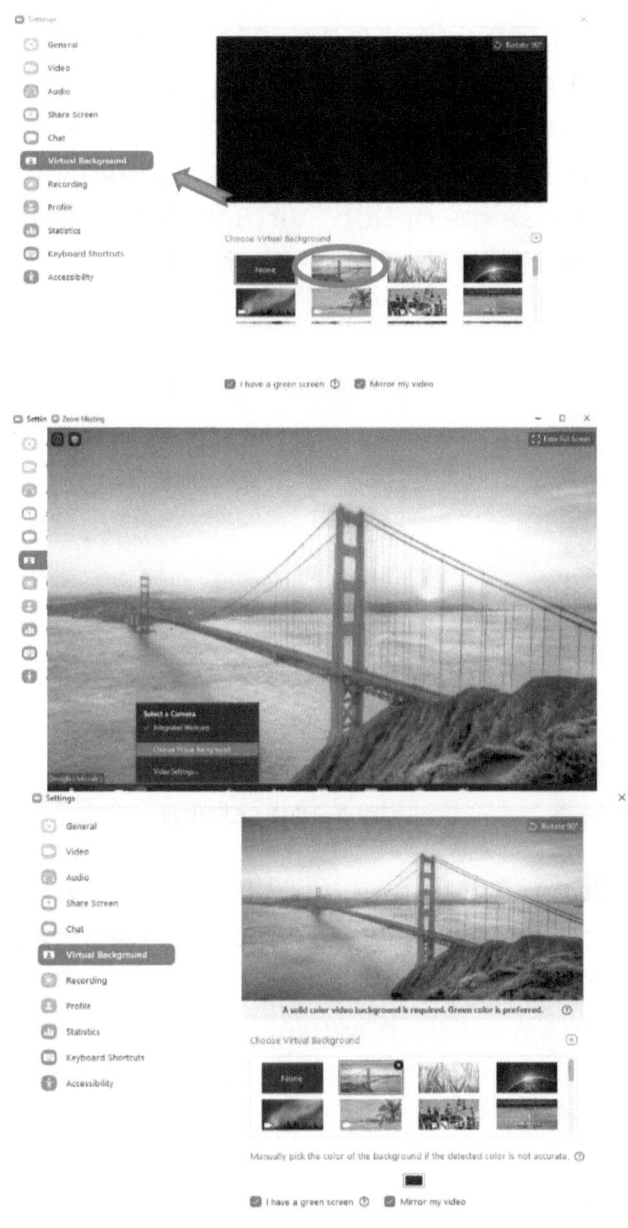

Desktop app screen

You can add the background you want on a meeting and change it as many times as you want, you just need on the arrow next to the video symbol and choose the option "Virtual Background" and repeat the previous steps.

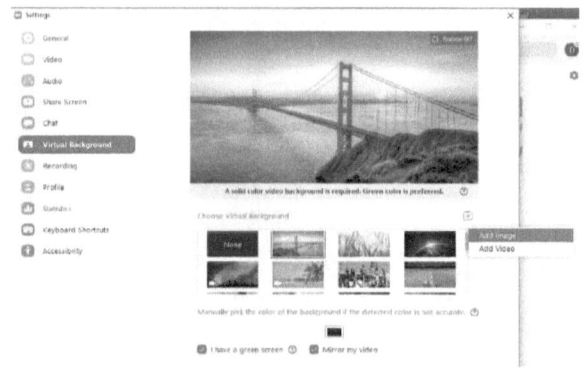

Desktop app screen.

Zoom recommends using a green screen and a good webcam to obtain the best results with virtual backgrounds, however, it is still possible to use virtual backgrounds without a green screen.

To use a virtual background in the mobile app, join or start a meeting, then touch the three dots that appear on right bottom corner, click on "more" and select the virtual background you want to choose

Zoom helps you to look better at a meeting with an option. Zoom allows to make a quick retouch to your appearance when you participate in a meeting so the other will see a shaper imagen with a filter that helps you look more natural.

You can activate by clicking on settings and in the video settings, click on appearance.

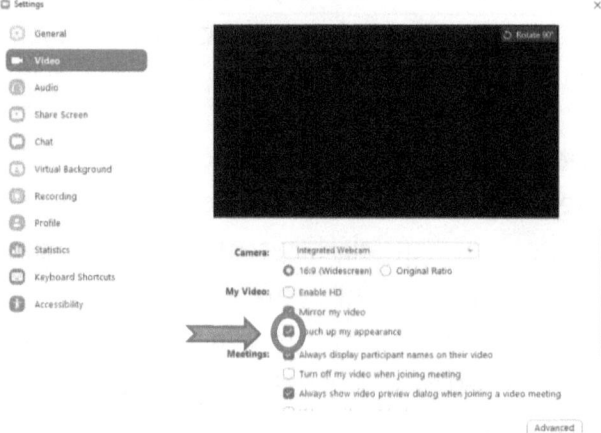

Desktop app screen.

4. Besides recording sessions, Zoom allows the automatic transcription of the audio in a meeting recorded in the cloud, as the meeting host, you have the possibility to edit the transcription, scan he text and search for keywords in the recording and share them.

This function allows is activated from Zoom website, click on the meeting configuration and in the recording option from the meeting, make sure the transcription option is activated. If it's not possible to do so, this may be because it is blocked for your account or group, in this case, refer to your Zoom administrator.

5. From gallery view, you can see up to 49 participants at the same time instead of the 25 pre determined, depending on your device.

From the mobile Zoom app, the pre set up configuration is to have the microphone on, if one or more participants join the meeting, you will see their small sized video on right bottom corner. From the mobile app you can see up to 4 people simultaneously per screen, if the number is more than 4, you will have to swipe on the screen to see the rest of them.

if you wish to see more and up to 49 participants per screen, you will need to join or start the meeting from the desktop app, once you are already there, go to settings and click on video to set the 'show up to 49 participants in gallery view'

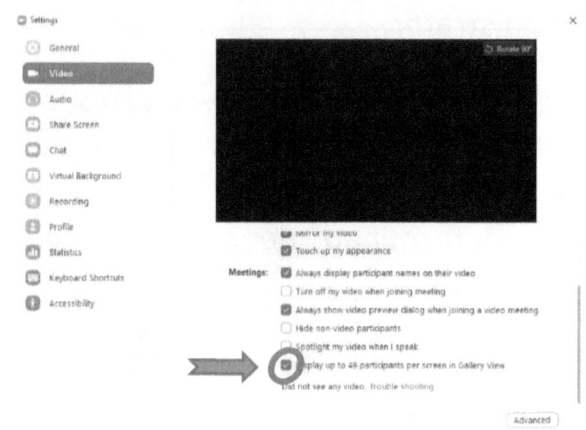

6. Zoom allows screen sharing whether it is from the desktop app or the mobile app, however, it also allows to pause screen sharing by simply clicking on "Pause" whenever there is something that you don't want to share with the participants of the meeting.

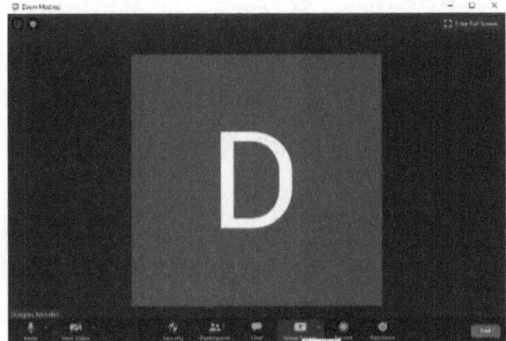

Desktop Screen App.

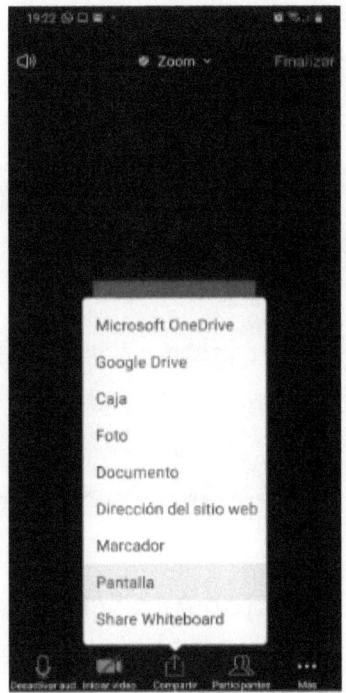

Smartphone Screen App.

7. Besides screen sharing, Zoom also allows files sharing as well as using the whiteboard to write comments anytime

On the other hand, it is also to write on the screen while someone is sharing their screen, to do this, click on options on the bar located at the top of the screen and then click on annotation, you will also find more options like text, draw, among others.

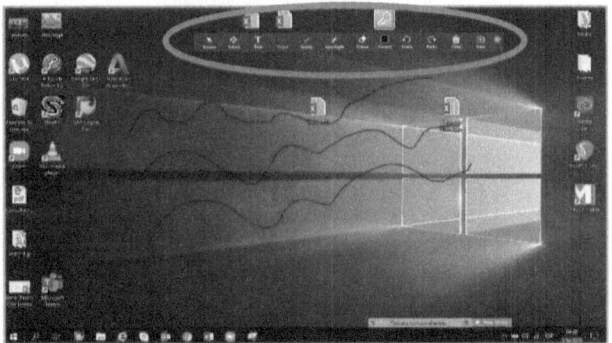

Desktop App Screen

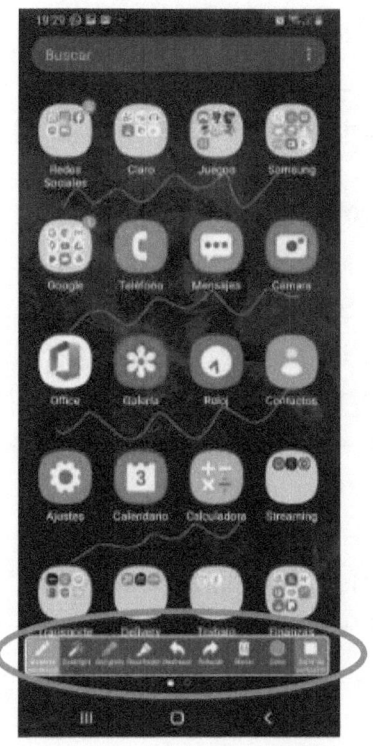

Smartphone App Screen

8. To make the meeting experience easier between the different tools and options that Zoom has to offer, there are some keyboard shortcuts and commands for the desktop app as well. You can find a list of them here.

- Alt + A or Command (⌘) + Shift + A = Mute / activate sound
- Alt + M or Command (⌘) + Control + M = Mute / activate audio for everyone except host
- Alt + S or Command (⌘) + Control + S = Start screen sharing
- Alt + R o Command (⌘) + Shift + R = Start / Stop local recording
- Alt + C o Command (⌘) + Shift + C = Start / cloud recording
- Alt + P o Command (⌘) + Shift + P = Pause o restart recording
- Alt + F1 o Command (⌘) + Shift + W = Change speaker's active view in a videoconference

We hope this guide helps you make the best out of Zoom and that your experience is much pleasant and you can enjoy videoconferencing in a better way, whether it is for comfort or work.